Sarah Howes has been a teacher and examiner since 1992. She has worked in a range of secondary schools and sixth form colleges and has variously been Head of a Maths & Science Faculty, Head of Biology & Psychology and an ICT Coordinator. She is currently Lead Consultant for Science in Surrey. Prior to teaching she worked in the NHS and Cancer Research.

Sarah is an experienced examiner for KS3 Science, GCSE Science and A level Biology and is a trained Standards Unit Coach. She is also a freelance editor and proof-reader.

1

Simple Scientific Slip-Ups

The definitive and hilarious guide for students, teachers and examiners on how not to answer exam questions and how to avoid misconceptions.

Sarah Howes

Paperback ISBN 1904312292
13 Digit ISBN 9781904312291
Published in the UK by MX Publishing
10 Kingfisher Close, Stanstead Abbotts, SG12 8LQ

INTRODUCTION

I have spent a good many years as an examiner for a variety of examination boards and papers, including SATs, GCSEs and A levels. It always amused me to read "student bloopers" in papers and magazines and, I must admit, there were occasions when I doubted the reality of some of those printed. However, I can say with all honesty that those printed here are the genuine article. I would spend hours toiling over exam papers in the Summer terms and then, with the onset of September, would provide my staff with something to amuse them in that long run-up to the much-awaited October half term.

There is no doubt that I am a pedant; I am the John Humphrys of the Scientific Teaching World, so some of these entries have caused me great pain. Worse still, I am freelance editor, so it has been incredibly difficult to deliberately misspell words and then have that glaring red or green underlining – which must be ignored!

So who is this book to appeal to and, what am I hoping to achieve with it?

I hope it will attract a diverse audience – scientists, teachers, students and the general public. I want it to provide amusement, but also, to serve as a learning experience for those out there who have yet to sit the dreaded exams and who may find some useful tips in these pages. Obviously, there are some parts which will only be intelligible to those who have a scientific background, but, on the whole, most of the errors are quite obvious. Those that aren't, if I'm honest, even I'm not sure what they mean!

The big question for me was: "How long should I carry on gathering my information before I actually turned it into something readable?" Well, I have currently just finished this year's consignment of SATs papers and A level marking, so I thought now was as good a time as any...so here it is.

I hope I can bring a shred of humour to your lives as well as an insight into how not to answer exam questions!

Author Note:

As an educator I am well aware of dyslexia, learning difficulties, SEN and other conditions associated with inability to write/express oneself. My intention is not to mock those students who may well be suffering from any of these conditions.

For RS – who brought joy to my life

Contents

One year the papers I marked came mostly from a Catholic school – or is that:

Chathlic, catolik, cathirlic, chocolic, cathic, catlic, cathlik, clicoth, cathic, cliciol, cholic, catholich, chathlik

I wondered why when they were so unsure, they actually bothered to try!

The transmission of HIV and AIDS

- People are more away of the disease.
- People are more bewared of the disease.
- Many people (females) can be carriers of the gene of HIV but don't suffer from AIDS. Moreover since AIDS is an alarming death case, less cases would be reported than the cases who are infected with HIV virus but is benign and frequent regular check ups.
- HIV is transmitted more easily than AIDS. People living in the same houses would catch the viruse easily if contraceptions not taken against viruse. Especially in 3rd world

countries where there is insufficient water to drink.

- HIV has no effect on humans.
- HIV is at present treatable. This is as it is knowable to be controlled to stop the disease.
- Spreading of the virus is very easy. It can be easily given and easily taken.
- The use of drugs has led to a decreased decline in activation of secondary infections
- People with AIDS can have their symptoms treated e.g. flu
- AIDS is incurable and the immune system is weaker than HIV inevitably you die sooner
- Not everyone how has HIV may now they have the disis as there are no ovius syntoms of the disis
- Secondary infections such as numonia
- Example of host cell of HIV – human fat cell
- Example of host cell for HIV – E.Coli/lung

AIDS is described as a retrovirus :

- because if it starts in one country there is a 50% chance for that virus to reach neighbouring countries as well as villages
- because its spreading. It hits built up areas
- because it is found to originate from the tobacco plant which has been around for many years

The reason why the number of people known to be infected with AIDS is likely to be greater than the number of AIDS cases reported....

- people with AIDS know they cannot be cured so there is no sense in reporting it
- because people are more sexually active and HIV is extracted this way
- once they get AIDS they cannot survive long so they are dead
- because some people e.g. African prostitutes are immune to the disease

Name one use of monoclonal antibodies

- Headache tablets
- Dipstick pregnancy
- Pregnancy skits

How does *Salmonella* lead to food poisoning?

- A person with salmonella may go to the bathroom to excrete faeces. Salmonella may end up on his hand and he may touch uncontinimated food

General Microbiology Questions

- There are too many cells to count which will result in anamolgus (anomalous) results
- During the experiment there was an anomalous witch was due to ......
- You spread the agar.
  You close the agar.
  You incubate in a refrigrator.
- You must ensure you work near a Bunsen flame near you to kill an bactria in the air

- It is used in argiculacular (agriculture)
- Sinsesants – (senescence)
- The samples were placed on a platalate
- The culture will have some type of respirating material in it
- The phospholipid bilayer acts as a cover up
- The phospholipid bilayer protects it from desiccation and from attack by pathgocytes
- The phospholipid bilayer acts as a disguise to confuse the macrophages
- The phospholipid bilayer has teeth to attach to other cells
- Antibiotics are produced industrially by fusion with a rapidly growing cancer cell
- The longer it takes to collect, the more viscose (viscous) it is
- Yogurt mad with the UHT milk would not have any live bacteria
- The hydrophilic heads attach to the T-helper cells. They are not scared of water.

- Penicillin also contributes to the growth and repair of internal muscles and cells that have been broken or injured. It has also been used on plants for experimental   purposes.
- The penicillin is extracted and can be purified by removing impurities
- Penicillin is one of the main pain reliefs available. It is in most antibiotics because microorganisms and bacteria are not yet immune to penicillin
- Penicillium produces penicillin as a secondary metabite (metabolite)
- As it is a fungus the penicillium is kept in the fermenting vessel and is fed on large amounts of garbage
- Penicillium aids in the production of antibiotics by growing penicillium in a bath fermentation
- The time taken for the population to double is known as the dublication time
- In order to find the number of bacteria you would use a turbometer
- The starter culture would contain the final results of another product

- The process of antibiotics starts from the production of penicillium fron fungus that branch on things
- From the creation of new antibiotics many people managed to survive deathfull illnesses
- Penicillin can be seen as a featury (feathery) structure over the surface of foods
- During the phase known as the leg (lag) phase, there will be no increase in the number of cells
- $0.1cm^3$ was added to a pen (Petri) dish
- The RNA uses the DNA to produce offspring
- The maltose concentration is initiatilly very high
- Using aspectic (aseptic) technique ....
- Ascomycota fungi reproduce asexually by means of specialised sex hyphae that use chemicals to detect the presence of one another. There are 2 strains – positive and negative and the positive meets the negative they mate and produce offspring.
- There is a sharp rise in the level of the glucose as the sucrose becomes more scares (scarce)

- Cellulase can make brewer's mash taste sweater

- Cellulase is used in sacrification i.e. in the production of more sugar

- It can be injected in the form of passive community

- Sterile a wire loop with a Bunsen fire

- The final stage is known as the death stage. At the end of this stage all the cells will be dead.

- Only one type of organism can grow. Other types are inhabited.

- The fermentation is done by itself without the need of a supervisor

- The deaf (death) phase

- The rate of growth is double as fast

- Antiobiotic – something that combats living organisms

- An antibiotic is a dummy version of a bacteria

- Gram +ve and Gram –ve would look different due to the staining of their thylakoids

- The number of cells stays the same – this could be the result of the cells not being able to find the glucose and lactose molecules
- Antibiotics fight pagocytes
- Some oval shapes will be died purple
- Penicillin prevents bacterial growth by pretending to be a white blood cell
- Gram –ve have a thin cell wall so the counterstain dies
- Too many bacteria cause overcrowdness
- They can derivided from fungi
- Hear, the bacteria would appear purple
- Antiobiotic is something that can be taken oral
- The bacteria would be pinky and purply
- Helpa-T cell
- Gram –ve has a cell wall made of urine
- Gram +ve appear purple whereas the others are colours
- The bacteria would appear as puple dots
- Structure of Tobacco Mosaic Virus – polyplegic
- Rhizopus feeds on dead living organisms

Pigments in plants and animals

- P_{FR} is slowly converted to P_R in dark light
- Cone cells are arranged in a one-one arrangement with the nephrons (nephrons are found in the kidney – not the eye!)
- In flowering animals……
- Pigments such as photochrome that are to blame for germination
- Pigments are small structures present in organelles such as chloroplasts
- Most pigments depict phototropism
- Animals detect light through their skin
- Main pigments in plants are chlorophyplast and carelanoids
- In the dark rhodopsin splits into rhodop and opsin – this is called regeneratoralion
- The pigment photofil is found in plants
- Rodopsin and iodopsin are exceptors of light
- UV light is dangerous and can produce skin cancer so mylin in the skin soaks it up
- These pigments allow animals to see and to be able to seek out danger

- Pigments are crushal to animals and plants
- Animals have evolved eyes where along the retina rods and cones detect light in the process of photosynthesis

Kidney

- The ascending limp/the longer the descending limp (limb)
- Distilled convoluted tubule
- The distal convulsed tubule
- Proximal distillated tubule
- So that the urine ends up being as concentrated and wasteless as possible
- ADH causes a greater concentration on amonia in urine
- More water is reabsoped
- In the kidney water is drawn out by photosynthesis
- The desserts water is highly valued as there is little available and mammals sweat and pant a lot
- The more areas the water has to reabsorb the more water will be, the further the loop of Henle pertudes into the medulla
- The pituarty galand/pititury gland/pitaurtary gland (I think you'll find that's Pituitary!)

- This stimulates the pastier (posterior) lobe of the perturity gland
- Mammals with long loops of Henle breed to produce long Henle offspring

Respiration

- Waste products of anaerobic respiration – carbon dioxide and abscisic acid
- Waste products of anaerobic respiration – Yeast extract (used in marmite)
- Final product of glycolysis – glycogen triphosphate
- The ATP is used in metalbolic activities

Nerve Impulses

- Speeds up impulses as they move by propergation
- Schawn cells/shaun cells/sacchwann cells/swan shells/swhann cells (Schwann)
- It is a Schwann cell that has raped itself around an axon

- It gives neurones that white colour you see in dissected animals
- The nodes of Varvier (Ranvier)
- The myelin sheet has Schwartz bodies
- Myelin sheath is a unicellular sheet of scumous cells
- Myelin sheath consists of schwan cells wrapped around the oxygen

What are lymphocytes removed from?

- The liver found in the spleen

Question on Ciliated epithelium & mucus production

- The mucus catches alien particles in the body
- It produces mucus which catches the unwanted being
- They secrete mucus which destroys anything which is alien in the lungs
- Smoking paralises the cilia for one day
- Produces mucus which kills nasty bacterea & has hairs which catch nasty bacterea
- This distroys the alvioli as it causes you to cough up the alvioli a lot
- The dirt and bad gases stick to the mucus
- The hairs through out mucus to clean the nose when we sneeze
- Secrete mucus which keeps the lungs clean of unwanted enzymes
- Smoking kills the alveoli and stops the bad systems from entering our bodies
- Mucus secreting cells emits tiny antibodies which stick to germs gobbling them up

- The hairs catch small particles such as tiny files

What useful substance does blood take in from the intestines?

Cells

Blood vessels

Liver

Heart

Brain

Digested food is absorbed into the blood from where?

Lungs

Excretion

Uterus

Intesties

Test tube

Anus

Umbilical cord

Question on How Artificial Respiration works

- Because if a person can't respire on his own, we would know what to put into him
- Because artificial respiration only requires carbon dioxide

Question on why we have double circulation

- Because it circulates twice through the body before being oxodised by the lungs
- The heart can partially carry on and offers support if the person has a heart attack
- Because the blood goes through the heart twice even though it could go through just once
- It is called double circulation because the gas goes through the body, heart and lungs and then passes back through them all again

Question on filling in the blanks

nose → trachea → _____ **→** _____ **→ alveoli**

lungs   liver

lungs   small intestine

lungs tracheorites

stimulus →____ **→ neurones →** ____ **→ response**

receptor   control unit

Enzyme	Action on Substrate
Fungal α - amylase	it makes the cake big and pluffy

Name a substance that affects nerve impulses across a junction

plastic; water; plasma; bone; stubbing your toe; electricity; earwax; shivering; oranophorous insecticides; tar; blood

Name the chemical messengers in the blood

chloroplasts;   xylem;   pathogens;   chromozomes;
gametes

Question on how alveoli are adapted to their function

- Alveoli have a large surface area - they are similar to roots
- They are adapted to the right size
- They can convert oxygen to carbon dioxide
- Alveoli have small root hairs which are very useful
- Alveoli have adapted to their job by ganging up on unwanted molecules and destroying them

Question on reflex actions and their importance

- They are instanius
- They keep the sensory cells healthy
- So we get minimum ammount or no pain when getting hurt
- Reactions to often fatal incidents would be slower and probably result in death
- As an impulse reaches the junction it goes through the dectonites

Question on Nitrogen/Carbon cycle

- When an animal is dead it lays on the ground
- The corps is absorbed back into the soils
- Dead animals and plants die every day which gives the carbon cycle something to feed on.
- Energy source for carbon cycle - carbon dioxide in the ground
- Dead animals and plants die...
- Dead plants and animals are dead.
- Living plants and animals die leaving dead plants and animals
- This means they begin to rot and when they do insects food on them
- When something dies its cells stay alive. The cells slowly use up the decaying object

Question on Inheritance of long-eye stalks in flies/Darwin

- A gene is a characteristic. Its what makes you have e.g. Down's syndrome on red hair
- The male had long eye stalks to impress the female
- Evolution takes place over a very long time. This is why it takes so long
- Short stalk has grown due to the female flies preferring them due to the different amount of sons being created
- Standard deviation occurs which led to long eye stalks becoming apparent
- Long stalks would help them see pray and prediters
- DNA is a gene
- Darwin observed that when left alone members of the same species will always mate and in fact this potential is never fully realised
- When mating the offspring may inherit the long eye stalk of its father (if it had one)

- A gene is something past down from your parents. For example you might have a gene from your dad which gives you a big nose
- The gene is the allele gene - this makes it what it is
- The long eye stalk is an obsessive gene. The short eye is recessive. This means that obsessive gene is likely to win
- The number of long stalks would increase because there would not be much point for the short stalked ones to live if the long stalked ones are more attractive
- Duration of prophase is very less

Question on why identical twins are identical

- Because they were born together
- Chromosomes will confront a substance that is strange to them and their alleles will be forced to stay the same
- Because two male gametes have been fertilised

- Each twin gets a replication of 1 DNA molecule because they come from the same parents
- Because they are made in the same embryo
- Identical twins have identical DNA because they have the genes
- They are born from the same egg
- Identical twins are made when the same egg and sperm fuse twice
- Identical twins have identical DNA because everything in the body is exactly the same for both of them
- Each person has 36 chromosomes which when fertilised can create out of 72 chromosomes which will mean that the new person will be unique
- Cross A produces more daughters than sons because the female chromosomes dominate the male chromosomes
- They come from the same eye

Question on Enzymes

- The molecules are more dense in boiling water
- With hot water lipase might boil and so it would release many enzymes before they would have time to work
- Lipase is immobile over the boiling point of water
- Hot water kills the fat
- If you boil the enzyme it gets too hot and melts
- Enzymes get killed after about 40°C (lipase) so the lipase does nothing as its dead
- The solution becomes acidic from being alkaline

Question on Photosynthesis, wavelength & filters

- The plant can only photosynthesise to a terminal rate

What colour filter would be best & why?

- Green, because it is the same colour as the leaf which affects the process and doesn't produce the required heat
- Green, because it would be red because it is the only chloroplast in the spectrum to make energy
- Green is the same colour as the plants themselves so it will have less of an impact

- The heat from the heater will move the air around, moving and shaking the plants
- More air can be defused
- It will rise the temperature and therefore it will cause it to dry up. It will then photosynthesise more so it can respire
- Hot air from the heater rises so the oxygen produced travels up through the window
- Too much light overwhelms the plant

- The warmth keeps the plant happy because it can photosynthesise better

Questions on Babies/Birth/Sex

How does a baby get oxygen while it's inside the uterus?

- *Babies don't breth until the umbilical cord is broken*
- *It breves water instead of air*
- *Babies don't breathe so oxygen isn't needed*
- *Through the philopian tube*

What is the normal length of pregnancy in humans?

- *16 months*
- *18 months*
- *36 months*

Why did the thickness of the lining of the uterus decrease between days 1 and 5?

- *The blood contained in the uterious is no longer needed due to already had intercourse*

Question on Osmosis, water potential & movement of substances into cells

- When the blocks were placed in the solution, the percentage change increased because the masses of before and after were different
- The change of mass was positive , otherwise it was negative
- The less concentrated the solution the smaller the sucrose atoms, so they could diffuse into the potato
- The sucrose solution was so strong that the potato dissolved and so lost a lot of mass
- The water stored in the blocks wanted to make the concentration the same
- The mass increased because the water wanted to go into the potato. This is because water went to places of less water than more

Questions on Ecology

- When they add limestone the stones trapped humanity and small vegetables start to grow around them
- Marine life in the forest would be destroyed
- When nitrogen fixingtion is add to the soil of the tree
- The stem started to be raptured due to the corrosive acid rain
- Acid rain damages the trees by deteriating it away
- The growth of some trees will be stumpted
- Growth will be stunned
- Sheeps/cows are allowed to graze on the field by people growing sheeps/cows
- Human activity is the main purpose of desertification
- Drought is caused by the grass being very flat so the clouds can go over them easily
- A lack of rain causes vegetation that isn't adapted to die
- The bark becomes flukey and falls off

- There is a constant pressure on dry grassland such as people trodding
- Overgrazing leads to a decrease in trees and stubbs
- The tree tastes a little sour
- Very less leaves are found
- The leaves will be moulded on
- Re-root a lake or small river
- Lack of birds and mammalian bodies within the trees
- Plant grass so the soil doesn't become dessert
- There are no trees left on any of the trees
- Humans overgraze the land
- Half the tree may be dead depending on which way the wind blew
- Therefore the land becomes more desertus
- So if conditions are scares like dry conditions or less green condition
- It's a viscous circle

Question on Greenhouse effect & the causes of more CO_2 in the air

- This causes the air to become carbonated
- It will see unusual temperature levels which will melt the ice gaps
- Humans uses cars more and giving off fumes. The fumes causes holes in the solar system making the sun shine through
- It will get hotter so it won't be cold in winter so animals may not hibernate when they are supposed to

Physics

The incident ray is brighter than the reflected ray. Give one reason for this.

This is because the incident ray is brighter than the reflected ray

What happened to the light rays when they hit the mirror

The lights went out

What forces act on the cup

Repeliation

A pupil plays her flute and the windows vibrate. Why?

The vibration is everywhere. It tackles the window

When the window vibrates, what happens to the laser beam that is reflecting off it?

- It will make a hole in it and melt
- It appears to be jogging

In 1781 scientists believed that there were 6 planets. Now they believe there are 10. What do these ideas suggest about our knowledge of the solar system

- We have found four more plants and technodge has improved
- We don't know nothing
- Its still not quite rite cause there are only 9
- We can't be sure for certain
- Scientists don't know what they are talking about

Suggest one way that developments in equipment have changed the information scientists collect about planets

By telegraph polls

What causes scientists to reject one idea and replace it with a new one?

Because some scientists are thicker than others

Why didn't the experiment work?

Because hes a knob

Why did the paper strip have to be removed from the torch before it would work?

So he did not get electriculed

Why is the circuit of torch A not complete?

- Because inside the bulb is something fishy
- The bulb has run out

What happened to the compass needles?

They stuck to the magnet

What do our eardrums do when sound reaches them?

- Bang
- Make nosies
- Eardrums make you hear what the sound waves say

Give one reason why we cannot hear properly when our ears contain a lot of wax

Because your ears are durtey. Take the wax out

Why do the wires need to be insulated?

To hide all the lively wires away from children

Friction is less on snow than on concrete. Give the reason for this

Because theres less friction

What are magnets made from

- Magnet
- lldm

What else protects parts such as doors from rusting

Wing mirrors

Give two substances needed for iron to rust

Get wet, be old

What is the name of the force that slows her down?

Drag act

General Biology Answers

- When decayed, plants produce a fluid called humerus
- Our reflexes act without our having to think about the action, which saves time when we are being burnt, stabbed etc
- Because they are insects they have infra-red vision
- Because the paint is sensible to daylight
- The leaves, petals and fruit are degayed by bacteria
- Dark banded snails can be camouflaged. However, an unbanded snail would look suspicious
- Because the light is not strong enough, this confuses the eye
- Nitrates are released when humus is broken down by sacrophytes
- What is filling space D? - nothingness
- Leaves fall to the ground. Some animals eat them and excrete them in their waist and the nitrogen in there phesies

- All the nutrients, vitamins and nitrates are presses into the ground and become the ground
- A reflex action is done ortomatic
- Tube D was boiled thus defacing the enzyme
- Pesticides are only used when pests are forecast
- He needs to repay his oxygen dept
- t is recessive because T is a killer gene
- How do reflex actions protect us? - slight uncomfortably is felt we do not have to wait for the damage to occur
- When fruit and petals reach the ground, they begin to compose
- Once the leave reach the ground it decomposes causing amonia and amoniom
- As they receive less water they do not become greatly turgor
- If we didn't have reflexes then as soon as the pain hits you it would be there quicker until you conciously thought about moving it
- Why does a camel have less wool under the body? this means that when the camel is too

hot the sweat can run off the wooly layer and drop off at the bottom

- Because they are less turgid which means they are flaccid
- Because the dark shells keep heat in and heat is attrcted to the dark
- When the leaves fall to the ground a series of sacrificial fungi and bacteria attack it.
- The enzyme(s) is boiled thus disenabling them to function
- X are specialised in preying on the mights. They will have specialised techniques to kill them
- It is recessive because it only produces one short combination
- Because being insects means they can find the mites quicker because of the food chain
- it is recessive - the one for shortness because with time they all just got faded out
- They will have better sensory organs adapted to finding there pray
- Why does a camel have webbed feet? it prevents the camel sinking in the dessert

- More water in bigger mammals means that its harder to lose equilibrium
- Denitrifying bacteria break down the faces
- What happens after exercising for long periods? crap begins to set in
- When food enters your windpipe you cough and wretch
- Explain why the mother is a carrier of this condition - because she carries the baby so people believe she is the carrier
- Recessive is when the alleles are different from each other
- Hairs act as obsitcals for the spiders
- Explain the graph - it is proportional on a line of best fit averaging that
- The short allele is recessive as it is present in the offspring but is overlapped by the dominant allele
- The lactic acid means what it reads like
- The gerbil would be a very large mammal, the camel very small. The man would be smaller than a gerbil but larger than a camel
- Because banded shells do not die of heat

- Starch is a saturated globular protein

Multiple Answers to Specific Questions

Alcohol causes the blood vessels in the umbilical cord to get narrow. Give one way this could affect the foetus.

Because the foetus wouldn't be able to collect food and drink from the womens ambicolic cord

An athlete can obtain energy more quickly by eating glucose than starch. Why?

- Glucose can stick to energy
- Because starch is fat
- Glucose has lots of protein which contains a lot of energy
- Glucose is a substance that overwhelms starch and makes it into glucose
- Because they are not attached it can be eaten a bit at a time

Give two effects of alcohol that would affect an athlete's performance

- They could make a fool of themselfs
- Loss of memory and calaps
- Keep on falling over

How does a scab protect the body

- It protects the germans not to go inside your blood
- It protects the bones from breaking and ripping the muscles
- It helps the cut not to bleed to death
- It keeps your body warm
- Because it would be hard and no longer an open womb

Why did they use a measuring cylinder

To stir with

How does white hair help an arctic fox to survive in the snow

- They may get colour blind in the snow and more can see them
- Yes
- It is undercover
- To show he is a block of ice
- To build an igloo

Why is hydrochloric acid needed in the stomach

- It is needed for your lungs
- To allow you to burp
- It helps you breathe
- To keep you alive
- So it can keep it morsturised
- To kill the food you eat
- The stomach is acid. It needs acid

In which part of the body are cilia found and what is their function there

- In the bottom part
- Middle women
- The root of the plant
- Eye - eyesite
- Skin – to clean your pores
- Near the ear
- Air cells
- Hair
- Leg
- Back
- Head – to stop your hair from falling out
- to control your brain

- Male testicals – to produce children to reach the ovarys
- Stomach – to help you think better
- Body – to absorb blood

Why could she remove paint with white spirit but not water

- Because white spirit is a chemical we can't drink and water we can
- Because white spirit is an alkali and is forced to take the paint away
- Because white spirit is the thing to use
- White spirit has a substance called parrafin water didn't
- No O_2 in H_2O
- Because she used white paint
- Because spirits has all kinds of liquid in it and water has no liquid in it
- Because it is antispirital

Explain why it is more difficult to compare the effects of drinking water on feeling more beautiful than on blood pressure

- If your ugly a few drops of water wont do anything. They don't make nivea visage for nothing
- Because the first statement is medical whereas the second is mental

How you would prepare a root tip squash

- It must be crushed using a pestil and morta as this doesn't damage the root
- Using a pick, stick the root tip
- Cover with the underslip
- Squash it using a hammer
- Using a mountain needle

How do lady birds get rid of aphids

- By scaring them with they're outside shell on there back
- They've got a special fluid in them
- By going to the toilet

How would this affect the number of foxes

- Yes
- Quite badly

Suggest one reason why farmers like to have barn owls on their land

There nice creatures

How can you tell the cell on the opposite page is from a leaf and not a root

It looks more complicated

The comet was last seen in 1986. Predict when it will next be seen.

1759

1770

1731

Which children are most likely to have freckles and how did you decide?

- I just guest
- Because they live in the same house and are more likely to catch freckles
- Because he is there son and has inherited there genses

Why is Yasmin's conclusion better than Harry's?

- Because she is a girl
- Becos its more persivic (specific)

There are 5 groups of animals with a backbone - 4 are shown above. Which one is missing?

Cannible

Dog

Hippo or gerroff

Fox

Carnivals

Omphids

How do spines help a porcupine to survive?

- Without them they couldn't get any air
- If someone wants to commit surside they cant cos it will hurt
- If they were shot it wouldn't travel through the spines
- No animal can eat them up unless they have hard lips
- It throws them at predators

Give the name of a gas which dissolves in rain water to give sulphuric acid

- Steam
- Petrol

Give one way in which a seal is adapted for moving through water

- It is slim line
- By swimming and sliming
- Because they have slimmy skin
- By its flaps

In what way are bats unusual mammals?

They are more like insects

When doing this investigation, why was it important that all the pupils jumped together?

Because they were so happy to be part of the investigation

Partridges lay their eggs in nests on the ground. How does this help them to survive?

There are lots of wild animals on trees

Why does a human egg not need to contain a food store for the embryo?

- Because they dunt after go threw birth
- It is because the person gives the embryo food by eating herself
- Humans are not made out of yolk
- The mother feed it babe
- Because we don't eat humans
- A cork which goes into the woman and egg
- The embryo gets its food supply from the mother via the urethra
- Through the embycal cord
- Through the embleical cord
- Through the philopion tube

Give one factor she could keep the same

She could change the size of the paperclip

How would Molly be able to tell if a more vigorous reaction had taken place?

Open her eyes

How could she improve the data she collects?

By trying harder

Give the name of the process by which gases spread through the air in a room

Gasification, acid rain, fertilisation, perspiration, particle dispension, germalisation, combustion, contamination, pollination

Write a scientific conclusion for the investigation

- The experiment was easy
- A measuring jug
- It was not that good

Suggest why there are no caterpillars on the tree in winter

- Because the wind would blow them off
- They are all keeping warm at home
- Because they are asleep in the kakuns (cocoons) to wake to be a butterfly

Why is eating too much fat bad for you?

- The fat gives you diriaer
- It makes you obeast
- It can rot your teeth

Give one way in which the trout is suited for moving in water

It can move by wagging its tail

Why is it an advantage for frogs to release large numbers of eggs and sperm?

- Because there a popular animal
- Because if they didn't they could blow up
- Because frog lay more eggs than human birds
- The frogs will wake up to many other frogs
- So that they can catch flys easy cause if they didn't the frog would be fat and it would be harder to move

How could Emily tell that the deer was male

Because they've got hoofs of the head

A red dye is added to the colourless alcohol used in thermometers. Suggest why

So people will like them more if they are alcoholics

What is the name of the process (weathering)

Expandiations

Which of the four nutrients, protein, carbohydrate, fat or calcium provides most of the energy in the cheese?

Banana

Complete the word equation:

Sulphur + Oxygen → BANG

How are substances carried around the body?

Down your mouth and into your bladder

How do the results show that maggots do not come from the air?

- Nothing can be born in the sky
- Maggots cant fly
- Because they always touch the floor

Which organ would be removed from the mouse during monoclonal antibody production?

- Blood
- Limbs

What could the owner put over the statue to protect it from acid rain?

An umbrella

....and instead of Biological Pest Control :

preditorisation; termination; extermination; unnatural prey; scare treatment; natural more economic control; insectidisation; slectlive breeding; humane;

for Synapse : cynapes; synaps; synethis
for bronchiole : broncholies; bronchol; bronchius; bronchials; bronicles
for emphysema : emphasyma; emphasmium - and also

Weird & Wonderful Spellings.....

Fungy
toung; tounge
liptease
bactria; bactiria; backterea
obsiticals
bille
shugar solution;
citroplasm
liptase; chlopyll; symbiotic fluid

Iodpisin

Rhoposin

Fitochromes

Ethanol acid

Answer continued on an additional sheath

A plato is reached

Monodisaccharide

To avoid contermination

Denaturd

Survivle

Increased awerness

Hydrolosysis; hydrolic (hydrolysis)

Taurpalling (tarpaulin)

Centrofuing

Dioreha; dioreah; diore; dierah; diorehha

Sarmonelar; Salomellar

Contrasption (contraception)

Homocide (killing all microorganisms)

Fallagleum (flagellum)

Indervigual (individual)

Neseccarry (necessary)

Visa vasa happens (vice versa)

Heart dissies

Chicking pots

Fomenter (thermometer)

Rong

Fervist (furthest)

Fing (thing)

Lhrvh (larva)

Reisdint

Upcrust

Firicsen

Erapsuin (eruption)

Vaireabal

Arrcuate

Exctriclty (electricity)

Hegroars (hedgerows)

Veichle emunitions

Measuring slinder

Loobrecates the joint

Centre meaters

More acurot & efishant

Generally Strange Answers!

- It can be harvested only ones
- The bacteria enter a lag faze
- Bacteria can puterfere (putrify)
- Enzymes are very small organisms
- Microorganisms are used in many uses
- The number of cells decreased more heavily drastically
- Lactic acid bacteria may be added to drive off the other naughty bacteria
- The steam increases the temputer
- The material is broken down into a slug-like matter
- Microbes respire producing ethanol as a waist product
- The number of cells is likely to be enormass
- Steam is added to kill any remaining microorganisms. This is necessary otherwise they would still be alive
- Some microorganisms produce oxygen from eating the sewage which is useful
- Steam is used to produce ATP

- This is done to sterile any microorganisms
- Lymphocytes are very sluggish and reproduce seldomly. Myeloma cells reproduce excellently
- If myeloma cells are used to produce hybridomas they become immoral
- The salmonella in the intestine secrets toxins
- The bacteria affect the pH and denature the enzymes catalysing the food
- This does not lower the pH as low therefore less is prevented longer
- The culture were customised to the incubator
- Describe the function of D – no
- They are buyest/biest/beast – biased
- Bilaball corde (umbilical cord)
- Explain your answer – not sure, I saw it on a film I think
- So she will have a nough for her and the baby
- The fox is a fety and the partridge a fay
- Skin colour can changed and weight can be lossed
- Because it is more serfistercated
- Its not in secwens (sequence)

- So they don't go exstinked
- Zinc is not a metal therefore it cannot magnate
- As a pose to
- Limestone inhabits the aluminium
- The graph flacurated
- So the bones can move apart when the agnostic muscels are at work
- Light contains vitamin C
-

....**and finally from a student who gained just 11/90**

Thanks for marking this disastrous paper – now go and make yourself a cup of tea – you deserve it!

The Three Minute Test

The biggest way students lose marks in exams is by not reading the instructions properly. This is a short test, but the instructions are very important and quite complicated. So it is important that you read the whole test **VERY CAREFULLY** before you attempt to answer any of the questions. You have only **THREE MINUTES** to complete the test and it is important that you get someone to time you.

ABCDEFGHIJKLMNOPQRSTVWXYZ

1. The above alphabet has one letter missing. Which is it?
2. Which letter comes three before R
3. Which letter is exactly between G and O?
4. Which the letter is two before the one that is halfway between Q and V
5. Which letter is three before the letter that is four letters before N?

6. Using the alphabet above turn the phrase "the cat sat on the rat" into a code by substituting for each letter the one two places further down the alphabet

7. What is the answer to the question before question 6?

8. What is the middle letter of the alphabet shown?

9. Which letter is four letters after the one that is five before K?

10. Now you have finished reading only answer question one

Acknowledgements

A book is never complete without the acknowledgements, but I have never liked those books which go over the top 'loving' everyone in sight!

My thanks, in the main, go to the students – without whom none of this would have been possible, to the exam boards I have worked for – Edexcel, AQA, OCR and before them, London & AEB, Steve Emecz, my friendly publisher and all those who have encouraged me over the years to put pen to paper and produce this book.

Other educational titles from MX Publishing:

Seeing Spells Achieving

Hickmott and Bendefy

ISBN 1904312209

Wonderful review in 2007
Summer British Dyslexia
Association magazine

Performance Strategies
for Musicians
David Buswell
ISBN 1904312225
*""If you suffer from stage
fright and performance
anxiety then help is at hand"*
The Pianist Magazine

Succeed In Sport

Jackie Wilkinson

ISBN 1904312241

"It's the sort of book I would have benefited from at the beginning of my sports life."

Sportsreach

Printed in the United Kingdom
by Lightning Source UK Ltd.
123382UK00001B/97-132/A